IN PROFILE

Women of the Air

David Mondey

SILVER BURDETT

In Profile

Women of the Air
Founders of Religions
Tyrants of the Twentieth Century
Leaders of the Russian Revolution
Pirates and Privateers
Great Press Barons
Explorers on the Nile
Women Prime Ministers
The Founders of America
The Cinema Greats
The War Poets
The First Men Round the World

First published in 1981 by
Wayland Publishers Ltd
49 Lansdowne Place, Hove
East Sussex BN3 1HF, England

© Copyright 1981 Wayland Publishers Ltd

Adapted and Published in the United States by
Silver Burdett Company, Morristown, N.J.

1982 Printing

ISBN 0-382-06634-0

Library of Congress Catalog Card No. 81-86277

Phototypeset by Direct Image, Hove, Sussex
Printed in the U.K. by Cripplegate Printing Co. Ltd.

Contents

Amelia Earhart — 5
  Amelia receives a hero's welcome home — 8
  Amelia flies alone across the Atlantic — 12
  Round-the-world trip ends in disaster — 16
  Dates and events — 19

Amy Johnson — 21
  Amy dreams of flying from England to Australia — 24
  The great adventure: Amy takes off for Australia — 27
  Amy takes off for the last time — 32
  Dates and events — 35

Jean Batten — 37
  Third time lucky for Jean — 39
  First woman to fly across the South Atlantic — 43
  New Zealand's heroine sets more records — 47
  Dates and events — 49

Jacqueline Cochran — 51
  From rags to riches: Jackie's early life — 53
  Jackie combines business and aviation — 55
  World acclaim for Jackie — 59
  Dates and events — 61

Glossary — 62
Further reading — 62
Index — 63

Amelia Earhart

> In June 1928 the name Amelia Earhart was on everybody's lips. 'First woman to cross the North Atlantic by air' proclaimed the newspapers. But Amelia was embarrassed by all the publicity as she had only been a passenger on that flight. She spent the rest of her flying career earning the praise of the public in her own right. Then in July 1937 her name once again hit the headlines . . . 'Pacific claims Earhart'.

In early June 1928, the inhabitants of Trepassey, Newfoundland, saw the arrival of an unusual aeroplane. This was a Fokker high-wing monoplane, powered by three engines, one mounted on each wing and one in the nose of the fuselage. This particular aeroplane was unusual because instead of conventional wheeled landing gear, it had two large floats which enabled it to land on and take off from water.

It had been acquired by a Mrs Frederick Guest, American wife of a British Air Ministry official, who planned to become the first woman to fly as passenger on a non-stop North Atlantic crossing. The aircraft had been given the name *Friendship*, for its journey from America to Britain was intended to strengthen Anglo-American relations. Wilmer Stultz had been selected to pilot the aircraft, with Louis Gordon as its flight mechanic, and Mrs Guest would go along as passenger. Apparently, this lady had second thoughts about this hazardous flight, and a little-known American woman, Amelia Earhart, was chosen to replace her. One of the reasons for choosing Amelia was the fact that she had learned to fly, and this meant that in an emergency she would be able to act as second pilot.

Unsuitable weather and minor mechanical problems delayed the moment of take-off. Shortly before 11.00 hours on 17th June, Stultz positioned

The Friendship *was fitted with two large floats before the transatlantic flight to enable it to take off from and land on water.*

the *Friendship* at one end of Trepassey harbour, facing the wind. The first attempt at take-off was a failure, as the heavy floats and extra fuel made it impossible to lift the aircraft from the surface. A second try was successful, despite waves that broke on the floats and showered the outboard engines with spray. The *Friendship* clawed into the air with both outboard engines spluttering in protest at their sea-water dousing. Fortunately, these conditions did not last long, and the aircraft was soon climbing slowly into a grey sky, heading east towards England.

Some 3,860 kilometres (2,400 miles) of ocean stretched between them and their intended destination of Southampton. It was to prove a nerve-wracking journey. Almost the entire route was blanketed in cloud, which meant that most of the flight was made on instruments alone. This was difficult enough, but was complicated by the fact that their radio had failed about nine hours after take-off. So it was impossible to get an accurate check on their position; Stultz had tried, unsuccessfully, to fly above or below the cloud. After climbing to 3,350 metres (11,000 feet) in a fruitless attempt to reach clear sky, he decided not to waste any more time or petrol.

Believing they must be nearing the coast of Ireland, and with only enough fuel for two more hours of flight, Stultz began a slow descent in an attempt to sight land. Lower and lower they went until suddenly, through a break in the cloud, they saw a large transatlantic liner. If only the radio had been working their problems could have been solved immediately. As it was, this ship only doubled their uncertainty. They expected it to be travelling along the same line as their aircraft, either on its way to or from America; instead, it was steaming almost at right angles to their line of flight.

Fame for Amelia

One can imagine the feelings of the three on board *Friendship*. Contrary to what their instruments showed, were they really flying more or less due north or south, rather than east? After a hurried discussion they decided they had no alternative but to have faith in their instruments, and continue on their original path.

Then, with their fuel almost gone, came the moment when they had to make a landing. Stultz headed *Friendship* down through the clouds. Were there hills or mountains in their flight path? No one could tell.

In fact, they were still over water, and soon after breaking through the cloud they saw small boats and then, at last, land. Within a few minutes they had found a suitable stretch of sheltered water, and touched down in a shower of spray after a flight of 20 hours 40 minutes. They discovered that they had overflown Ireland and were off the coast of Wales.

On the following day the *Friendship* was refuelled and flown to an official reception at Southampton. Strangely it was Amelia Earhart who received most of the acclaim and subsequent fame.

Amelia stands beside the Friendship *on the quayside at Burry Port, South Wales after her epic flight.*

Amelia receives a hero's welcome home

Amelia comes home to a hero's welcome after transatlantic flight ... Writes a book about the flight under the guidance of publisher, George Putnam ... Early life and first interest in flying ... Works with pioneering airlines ... Sets an altitude record in a Pitcairn autogiro ... Comes third in America's first air race for women ... Breaks women's international speed record twice ... Marriage.

When she had first been approached to take part in the North Atlantic adventure, Amelia had expected to share some of the pilot's duties. But the weather conditions had prevented this, because at that time she was not experienced in instrument flight. She was very disappointed, and as her role in this crossing was little more than that of a passenger, she did not feel she deserved the acclaim of the press and radio. Stultz, as the successful pilot of a difficult flight, deserved the praise, but this was not to be. The newsmen had already exhausted their stock of praise for earlier transatlantic flights by male pilots. Amelia was the first woman to dare to attempt such a flight, and, passenger or not, she was given the full glare of publicity.

European excitement over this event, plus the riotous reception she received on her return to New York, left Amelia almost breathless. Her first task after the ballyhoo of the homecoming had died down was to write a book, entitled *20 hrs 40 min*. The idea of writing a book about the flight had been planned before she set out, for Amelia had already shown some promise as a writer. George Palmer Putnam, of the famous Putnam publishing company, had managed to persuade Charles Lindbergh to write the story of his solo flight, and had counted on Mrs Guest

As the first woman to complete a transatlantic flight, Amelia received much press acclaim.

The Avro Avian III which Amelia bought while in Britain after her first transatlantic crossing.

to do the same. When she decided to withdraw, George Putnam persuaded Amelia to keep a log so that she could write a book on her return.

Born at Atchison, Kansas, on 24th July 1898, Amelia was the daughter of a much-travelled railroad (railway) official, and so she was used to travelling from an early age. She was very much a tomboy and keen on strenuous games. At the age of twenty she gave up her studies at college, so that she could help the wounded servicemen of the First World War, who were then returning to hospitals in Canada. It was in Toronto that she first saw aeroplanes in large numbers, and her own interest in flying dated from that time.

Firstly, though, she took a course in motor car repair and maintenance. This was followed by studies to become a doctor, but despite obtaining successful results in medical examinations, Amelia realized that this could not be her life's work. Aviation was calling, and encouraged by her mother,

George Putnam is seen here leaning against the open cockpit of one of Amelia's aeroplanes.

Amelia began flight training. She mastered the basics quickly enough, and was soon able to fly solo. At that time it did not occur to her that she might earn a living in aviation. Instead, she drifted from one job to another, uncertain of her true path. Still concerned for others who were less secure, she finally went to Denison House, a thriving social centre in Boston, Massachussetts. There, she found great happiness in her work, and maintained her interest in aviation as a member of the local branch of the National Aeronautic Association. It was through her membership of this Association that the invitation came for the transatlantic flight.

Amelia's early flying activities

When all the excitement of the Atlantic crossing had died down, Amelia became unsettled by the longing to fly again. She soon turned this longing into reality by making a transcontinental flight in an Avro Avian III, which she had acquired when she was in Britain. She then worked as aviation editor of *Cosmopolitan* magazine, and later became involved with two of America's pioneering airlines, largely in a public relations role. This period was to set the pattern of her future life, and she enjoyed every moment of those early years of aviation. Civil airlines had begun to expand rapidly, sparked off by the trail-blazing flights of the pioneers, especially Charles Lindbergh's solo flight from New York to Paris in May 1927.

One of Amelia's early aviation activities was a demonstration tour in a Pitcairn autogiro. This type of aircraft is basically a conventional aeroplane. But instead of normal fixed wings, it has a rotary wing, which is kept turning as the aircraft moves through the air. It was in this aircraft that she set an altitude record of 5,613 metres (18,415 feet). Then, in 1929, with George Putnam's backing, she competed in

America's first air race for women. Not surprisingly, this was reported in the press as 'The Powder Puff Derby', but whatever the name, Amelia was happy to be involved and succeeded in gaining third place.

In the following year she attempted to set a women's international speed record. Flying a Lockheed Vega high-wing monoplane, Amelia set two new speed records at Detroit, Michigan, on 25th June 1930. Once again, George Putnam had been involved with the arrangements for these record attempts. On 7th February 1931, he and Amelia were married. From then on he took care of all their activities on the ground, leaving Amelia free to fly.

Amelia set a new altitude record in this Pitcairn autogiro.

Amelia flies alone across the Atlantic

Amelia standing under the wing of the Lockheed Vega monoplane before her solo transatlantic crossing.

Amelia's plan to fly solo across the Atlantic... Preparations for the flight... Take-off from Newfoundland... Flight dogged by problems... Altimeter breaks down... Thunderstorm gives way to dense cloud... Attempts to fly above cloud, but ice forms on wings... Flames break through the exhaust system... Fuel leak in reserve tank... Safe landing near Londonderry.

Amelia was now able to concentrate on flying. Her first task was to earn the praise in her own right for the achievement which had set her on the road to international fame. She had always been very conscious of having gained acclaim for that first flight across the Atlantic: praise which she considered was really due to Wilmer Stultz.

With certain modifications, her Lockheed Vega monoplane would be ideal for the flight. Amelia arranged with a friend, Bernt Balchen, to give the aeroplane a complete overhaul. Balchen was himself a famous pilot: among other achievements he had been the first to fly an aircraft over the South Pole, on 29th November 1929.

Balchen's work on the Vega included some reinforcement of the fuselage, because the passenger cabin was to hold a large auxiliary fuel tank that would enable the aircraft to fly some 5,150 kilometres (3,200 miles). Additional instruments were fitted to enable the plane to fly 'blind', and three compasses ensured that Amelia would make no mistake over her direction. Finally, a new engine was installed, and after initial test flights by Balchen, Amelia made frequent flights in the plane. During these sessions in the Vega, she concentrated on practising instrument flight, until she felt confident that she could fly to Paris in darkness or bad weather.

Take-off from Newfoundland

On 19th May 1932, Balchen flew the Vega to Harbor Grace, Newfoundland, the chosen take-off point for the Atlantic crossing. Amelia, and mechanic Eddie Gorski, travelled as passengers on the cabin floor. The take-off had been planned for the evening, so that Amelia would be rested and fully alert for the stage that had to be flown during the hours of darkness. After a reasonably promising weather report, the take-off was made at 19.12 hours on 20th May 1932. This happened to be the fifth anniversary of Charles Lindbergh's take-off from Paris.

Profile of a Lockheed Vega monoplane, similar to the one Amelia piloted on her transatlantic flight.

The first hours of the flight were easy, calming Amelia's excited nerves as she left the sunset behind and saw the moon rising. Suddenly she realized that the altimeter (height recording instrument) was no longer giving accurate readings. For the rest of the flight she could not be sure of the Vega's height above the sea. This presented few problems in good visibility, but by midnight she was flying through a thunderstorm, relying on instruments to maintain her course and height. After about an hour, she flew out of the storm, only to enter dense cloud. As Amelia climbed to get above the fog-like mass, she ran into ice. As the aircraft's rate of climb began to fall, due to the weight of ice on the wings, she was forced to descend into warmer air. Eventually she could see waves, but without an accurate altimeter she could not risk flying so low. She had to climb back into the cloud and fly just below the icing level.

Dawn came as an enormous relief, for it made it much easier to judge heights. Another problem had developed, getting steadily worse throughout the

Amelia flew from Harbor Grace in Newfoundland to Londonderry in Ireland in just under 13½ hours.

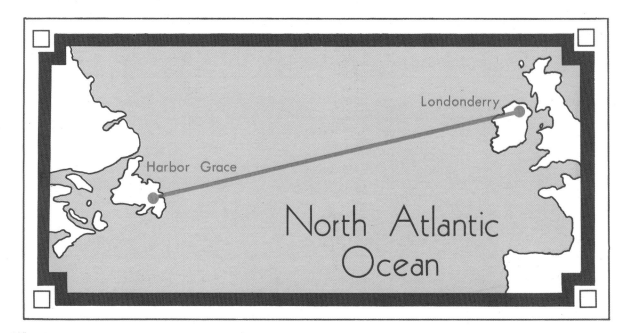

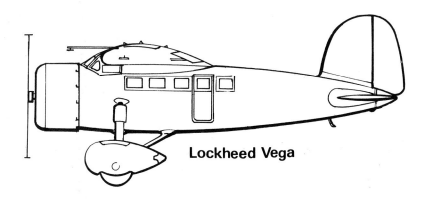

Lockheed Vega

After her solo Atlantic crossing, Amelia was affectionately nicknamed 'Lady Lindy' after Charles Lindbergh, the famous aviator.

night. About four hours after take-off flames had started to force their way through a crack in the engine's exhaust system. Now it was clear to Amelia that, before very much longer, there was a distinct possibility that a section of the exhaust system would break away, possibly causing considerable damage.

Landfall over Ireland

Then, when she turned the fuel cocks to bring the reserve fuel tank into use, she discovered that there was a fuel leak. The combination of all these factors meant she had to find a landing field as soon as possible. She abandoned her original plan of flying to Paris, and began to scan the route ahead for a first sight of Ireland.

Soon after 08.00 hours on 21st May, Amelia saw coastal vessels below her and very soon after, made a landfall almost exactly on target, at just about the centre of Ireland. Here she had planned to turn south and head for Paris: instead she turned north into better weather and, after circling a town without finding any sign of an airfield, landed in a meadow. The town was Londonderry, the time just after 08.40 hours. Amelia Earhart had completed the first solo crossing of the North Atlantic by a woman, in almost exactly 13½ hours.

Round-the-world trip ends in disaster

Amelia receives honours in America and Europe... Establishes new non-stop transcontinental record... More records... Plans for round-the-world flight... Take-off accompanied by Fred Noonan... First stages of journey uneventful... Final stage involves difficult landing on tiny island in the Pacific... U.S. Navy on hand to assist... Radio call received... Nothing more heard of Amelia.

Once again Amelia's name was on everyone's lips. One journalist, noting her striking resemblance to Charles Lindbergh, referred to her as 'Lady Lindy', a name she never lost. Amelia received honours in both Europe and America. Shortly after her return home she established a new national record. On 25th August, flying the Lockheed Vega again, she set a time of 19 hours 5 minutes for a non-stop transcontinental flight from Los Angeles, California, to Newark, New Jersey. This was a new solo record for a woman pilot.

There followed a quiet period until, once again, Amelia's name hit the headlines as she achieved a double first in the Lockheed Vega. This time she had flown solo from Wheeler Field, a U.S. Army Air Corps base on the Hawaiian island of Oahu, landing at Oakland, California, almost at noon on 12th January 1935. In a flight of 18 hours 15 minutes she had not only become the first woman to make a transpacific flight from Hawaii to the U.S.A., but was also the first person of either sex to accomplish a solo flight over the route.

Little more than three months later, on 19th April

Edith Benson celebrates with Amelia after winning the 1935 Earhart trophy race.

Amelia's Lockheed Electra airliner was specially equipped to enable it to fly round the world.

1935, she flew solo non-stop from Los Angeles, California, to Mexico City. On 8th May she made the reverse transcontinental flight. This time she landed at Newark, New Jersey, becoming the only woman to have made north/south, south/north non-stop solo flights across America.

This achievement was followed by a happy year as vocational counsellor for the women students at Perdue University. The trustees were so pleased with her work that they established a special fund for aeronautical research, enabling Amelia to acquire a new Lockheed Electra airliner. Equipped with a long-range fuel system and advanced navigational equipment, such an aircraft would be able to fly around the world, and Amelia soon began to make plans for this adventure.

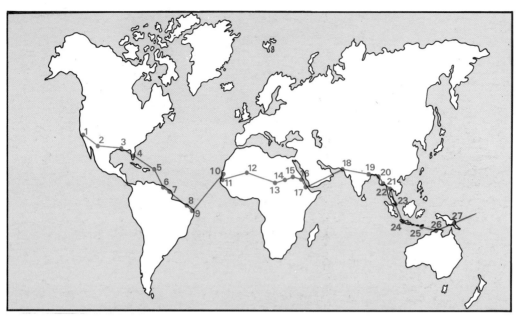

The route Amelia took on her round-the-world flight before she disappeared over the Pacific; 1 Oakland; 2 Tucson; 3 New Orleans; 4 Miami; 5 San Juan; 6 Caripito; 7 Paramibo; 8 Fortaleza; 9 Natal; 10 St Louis; 11 Dakar; 12 Gao; 13 Fort Lamy (now Ndjamena); 14 El Fasher; 15 Khartoum (now El Khartum); 16 Massawa (now Mitsiwa); 17 Assab; 18 Karachi; 19 Calcutta; 20 Akyab; 21 Rangoon; 22 Bangkok; 23 Singapore; 24 Bandoeng; 25 Kupang; 26 Port Darwin; 27 Lae.

On 20th May 1937, Amelia lifted the Electra off the airfield at Oakland, California, on the first stage of this round-the-world flight. It was not intended as a high-speed dash, but a leisurely first flight around the world by a woman pilot. Accompanied by Fred Noonan, a highly experienced navigator, she reached Miami, Florida, on 23rd May. The aircraft remained there until 1st June while servicing and adjustments were completed. Then the flight started in earnest, down South America to Natal, Brazil, for the hop across the South Atlantic to Senegal. On across Africa and India, then dropping down via Burma, Singapore and Java to Koepang, the traditional take-off point for Port Darwin, Australia. By 1st July the Electra was at Lae, New Guinea, and on the following day Amelia took off at 10.30 hours on the most difficult stage of the journey. This was to the tiny Howland Island in the Pacific, which was used as a stepping-stone to Honolulu, Hawaii, before take-off for America and home.

This was thought to be the most dangerous stage of the entire flight, as Howland was just a pinpoint in

To this day no one really knows what happened to Amelia Earhart on her last flight.

the vast Pacific Ocean. The U.S. Navy had positioned the USS *Ontario* midway between Lae and Howland. The coastguard cutter *Itasca* lay off Howland to assist with direction finding, and the USS *Swan* was between Howland and Honolulu. Radio communication was maintained between the *Electra* and the *Itasca* from 02.48 hours on 3rd July, but messages received from 06.15 hours indicated that Amelia and Noonan did not know their position in relation to Howland. After a series of radio contacts, when it became clear that they were low on fuel, a last call from Amelia came at 08.55 hours, giving only the aircraft's heading. Nothing more was heard from them, no wreckage found, no trace of their bodies.

In America, and soon around the world, headlines shouted 'Lady Lindy Lost' and 'The Pacific claims Earhart'. The topic of her disappearance was on everyone's lips, and the mystery is discussed to this day. A number of strange theories have been put forward to try and explain the loss of these two people, but no one really knows what happened.

Dates and events

1898	Amelia Earhart born at Atchison, Kansas, U.S.A. on 24th July.
1920	Amelia learns to fly.
1921	Makes her first solo flight.
1928	As a passenger, Amelia becomes the first woman to cross the North Atlantic by air (17th-18th June).
1930	Establishes women's international speed record over 100 km course.
1932	Becomes the first woman to fly the North Atlantic solo, east to west (20th-21st May).
1932	Establishes women's solo U.S. transcontinental flight record.
1933	Establishes new women's solo U.S. transcontinental record.
1935	Makes the first solo transpacific flight between Hawaii and U.S.A. (11th-12th January).
1935	Becomes the first woman to fly between Los Angeles and Mexico City (19th April).
1935	Makes first women's solo flight between Mexico City and Newark, New Jersey (8th May).
1937	Beginning of round-the-world flight (20th May).
1937	Amelia disappears.

Amy Johnson

One grey day in January 1941 Amy Johnson took off on a routine flight. But this time she did not return. Her mysterious disappearance has made Amy Johnson into an almost legendary figure. Amy had always been conscious of competing in a man's world. She was determined that men should recognize her contribution to aviation. Thus was born her ambition to become the first woman to fly solo from England to Australia.

Much of the story of Amy Johnson's final journey has to be pieced together from a few isolated facts. But we do know the events leading up to her mysterious disappearance. On 3rd January 1941 Flight Officer Amy Johnson received instructions to deliver a de Havilland Tiger Moth from Hatfield to Prestwick, Scotland. She was part of the Air Transport Auxiliary (A.T.A.) who carried out the important work of ferrying military aircraft between manufacturers, maintenance units and operational stations during the Second World War. It was a bitterly cold winter's day, and when she arrived at Prestwick, she was happy to learn that a message had been received requesting her to stay overnight. Instead of a tiring return journey sitting on a suitcase in an unheated corridor of a wartime train, she was to fly an Airspeed Oxford down to Kidlington, Oxfordshire, on the following morning.

The first stage of the return journey was quite routine. It was a grey, cold day and the cloud was thickening as she landed at Blackpool to refuel. When she learned that the weather was even worse further south, she decided to stay overnight. By morning the sky had cleared. The Oxford trundled across the airfield on the final stage of the journey as it had done so many times before. The aircraft's tanks contained enough fuel for about five hours' flight, more than sufficient to reach Kidlington without stopping to refuel again.

The rest of the story is largely guesswork. As the Oxford droned southward the clouds thickened and the weather became steadily worse. There seems little doubt that Amy, believing that she would be able to fly out of the clouds into clearer weather, decided to keep going, maintaining direction and level flight on instruments alone. Eventually she must have veered eastward of her flight path: the instruments may have been faulty, or perhaps there appeared to be less dense cloud in that direction.

Way below her a wartime coastal convoy was heading out of the Thames estuary, bound for one of Britain's northern ports. Suddenly, and almost silently, an aircraft broke through the low cloud, passing just ahead of the destroyer HMS *Berkeley*. 'Aircraft off the port bow', shouted the bridge lookout, and as eyes turned in response to his warning, a parachute could just be seen behind the aeroplane.

Amy's Airspeed Oxford passed just ahead of the destroyer HMS Berkeley *before it crashed into the sea.*

Despite a dramatic rescue attempt and a thorough search of the crash area, Amy's body was never recovered.

Driving sleet made it almost impossible to get a clear view. No one was able to identify the aircraft positively as an Oxford, for within seconds of that first brief sighting it had sunk. Even before the figure below the parachute had touched the icy water, the *Berkeley* was racing towards it as engines surged in response to the command 'full ahead, both'. Within a few minutes a rescue boat had been launched. A second escort, HMS *Haslemere*, was also heading towards the figure struggling for life in the swirling water. Seeing that the rescue boat would not make contact in time its Captain, Lieutenant Commander Fletcher, dived overboard to try and save the pilot. In worsening visibility, the ships' crews were unable to see anything. Then, when a sudden squall of sleet cleared, only the Captain could be seen in the water. When the rescue boat pulled him from the water he was almost unconscious, suffering from exposure. He collapsed and died without saying a word.

World heroine lost at sea

Although vessels of the convoy searched until darkness, no trace of the pilot's body was found, nor was it recovered later. Most of the seamen who had seen the event, or heard the calls for help, thought the pilot was a woman. A flight bag was recovered, containing an A.T.A. Flight Authorization form, but the seawater had made the inked-in details unreadable. It was later established that the aircraft which had dived into the sea was an Airspeed Oxford. No other A.T.A. pilots were missing on that day, and no wreckage of another Oxford with a woman pilot on board was reported or found.

But why should this pilot be picked out for special mention? Simply because Amy Johnson had become a world heroine in 1930 as the first woman to fly alone from England to Australia. Her life story is one of courage and determination.

Amy dreams of flying from England to Australia

Amy's early life and family background ... School and university ... First interest in flying ... Saves up to pay for flying lessons ... First solo flight ... Gains flying qualifications ... Ambition to become first woman to fly solo from England to Australia ... Looking for sponsorship ... Lord Wakefield helps ... Amy buys a secondhand Gipsy Moth biplane ... Plans for the flight.

All the women in this book were determined to succeed in their chosen field, and this was especially true of Amy Johnson. In the early 1930s, she not only set out to achieve a pioneering flight, but also to break into and compete in what was then a man's world.

Much of Amy's determination must have come from her Danish ancestry. Her grandfather, Anders Jorgensen, ran away from home to sail the seven seas. He boarded a fishing smack for the first stage of his voyage, which carried him to England. He liked the Yorkshire countryside so much that he decided to stay there, eventually starting a business as a trawl fish merchant and exporter, and adopting Johnson as an anglicized version of his name. Three of his sons inherited their father's business. One of these was Amy's father.

Amy was born in 1903, and like many of her friends became fascinated by aeroplanes while she was at school. It was not until after the First World War, in 1920, that she had her first flight as a passenger, spending the then large sum of five shillings (25p) on her fare. So far as she was concerned, it was a disappointing experience and a waste of money. It was not until some eight years later, with school and a university degree behind her, that the urge to fly returned. By then she was working in London, sharing a bed-sitting room not far from Stag Lane Aerodrome at Edgware. There she joined the London Aeroplane Club, one of the earliest private flying clubs to be formed in Britain. By economizing on clothes, walking to and from work, and eating only the minimum amount of food, she was able to save enough money to pay for one flying lesson each fortnight. These began in September 1928. She made her first solo flight at the beginning of June 1929, and gained the coveted private pilot's 'A' licence towards the end of that month.

By then Amy had decided to make her career in

This portrait of Amy was taken when she was about eight years old.

aviation, and immediately began to learn everything she could about aircraft, their construction, the engines that power them, and how to maintain the complete aeroplane. In due course she gained the ground engineer's 'A' and 'C' licences and, later, her navigator's licence.

Competing in a man's world

But she was still a woman in a man's world. She knew that she must achieve some startling success to prove beyond doubt that in the air she could do anything they could. On the ground she was already master, for very few male pilots were certificated ground engineers. The idea of making a solo flight to

Amy overhauling her plane whilst training to become an engineer.

25

Amy was determined to prove that women had just as much ability as men when it came to flying.

Australia was born at this time. If she could beat the Australian Bert Hinkler's time for the solo England-Australia flight, men would have to acknowledge her ability. By such pioneering flights, other women would in time be able to make a future for themselves in aviation.

It seemed an eternity before she could put her dream into practice, but in fact she managed to find the backing she needed more quickly than expected. This was because the people to whom she turned for help were already impressed by her determination. An appeal to Air Vice-Marshal Sir Sefton Brancker, then Director-General of British civil aviation, gained an introduction to Lord Wakefield, who had helped so many in their attempts to gain air, land, and water records. Lord Wakefield promised to help with fuel supplies, and gave a lump sum towards the purchase of a secondhand de Havilland Gipsy Moth biplane. Amy's father provided the balance of £350, and suddenly the project was within the bounds of possibility. If determination alone could get Amy and her little aircraft to Australia, she was already there.

The de Havilland Gipsy Moth in which Amy was going to fly to Australia.

The great adventure: Amy takes off for Australia

Amy and Jason *take off from Croydon aerodrome... Early stages of the journey... Flies into violent sandstorm... Engine splutters and cuts out... Engine-off landing in the desert... Wind drops and Amy is able to take off again... More problems with the weather...* Jason's *wing and propeller damaged... Emergency repairs... Final long stage across the Timor Sea... Lands safely.*

On 5th May 1930, a few minutes before 08.00 hours, Amy's Gipsy Moth stood ready for take-off at Croydon aerodrome. Its fuel tanks were filled to the brim; the forward cockpit packed with tools, spares, first-aid kits, essential clothes, and a variety of odds and ends; and a spare propeller was lashed to the port side of the fuselage. The little aircraft—named *Jason* after Amy's father's company—had been well prepared for the flight by C. H. Humphreys, the London Aero Club's chief engineer.

After a kiss for her father, and a brief farewell to the handful of friends that had gathered to see her off, Amy opened the throttle and *Jason* began its run across the grass. Within seconds Amy realized she had not allowed sufficient take-off run for the heavily-laden Moth. Quickly she closed the throttle and taxied back to take a longer run. The second attempt was successful, and accompanied by five Moths from the Aero Club Amy climbed slowly away to the south. At the coast the escort dipped in salute and turned for home. *Jason*, swallowed up in the mist and cloud that was still hanging over the English Channel, disappeared from sight.

This was a supreme moment for Amy. All the months of hard work, planning, fears and frustrations were behind her. As *Jason* broke through the cloud into sun and blue skies, it seemed almost like the beginning of a holiday. Ahead lay some 16,000 kilometres (10,000 miles) of adventure. Her training and dedication to aviation enabled Amy to face this journey with confidence and keen anticipation. At about 18.00 hours that evening she closed the throttle for the first time since take-off. *Jason* bumped across the turf at Vienna's Aspern aerodrome and came to a stop. It was almost an anti-climax—it had been so easy. Vienna to Constantinople (now Istanbul) and Constantinople to Aleppo presented few problems, but the first trial of strength was soon to come.

Jason *stands ready for take-off at Croydon aerodrome.*

It was a hot and humid day as Amy approached her next stopping point at Baghdad. A dark cloud ahead proved to be a funnel of sand, spiralling up from the heat and sandstorm below. *Jason* pitched and bucked and then, with a sudden cough, the engine stopped. Its roar was replaced by the howl of the storm. Wind screamed through the rigging wires, as the little craft plunged towards the desert below. Fighting with controls that seemed to have little effect on the aeroplane, Amy made an almost miraculous engine-off landing on the desert. Immediately she had to jump from the cockpit and hang on to *Jason* to prevent the aircraft from being blown over onto its back and wrecked. A lull in the storm enabled her to wedge luggage in front of the wheels. Then followed a grim, sand-choking two hours of hanging on until the storm abated. In her right hand Amy clutched a revolver for protection, in case any of the nomadic Arabs should try to attack her.

After what seemed an eternity the wind dropped. Would it be possible to restart the engine and take off? Amy pulled the propeller over several times before, almost reluctantly, the engine spluttered into life. She put the luggage aboard, and then managed to get *Jason* moving slowly across the loose surface, creating her own sandstorm as the little Moth climbed and headed towards the minarets of Baghdad.

Problems en route

The route ahead was highlighted by events that would have made a less determined person give up in despair. On the stretch from Baghdad to Bandar Abbas, a landing-gear leg collapsed as *Jason* touched down. When Amy landed at Karachi, she was two days ahead of Bert Hinkler's record, and the world was agog for news of her progress. She was forced down at Jahnsi, with fuel tanks almost empty after bucking a headwind. As she landed on the playing field of an army depot, there came a splintering crash as a wing collided with a building. Was this the end?

Amy hangs on to Jason *to prevent it from being blown over in a sandstorm.*

Jason had to be repaired after crashing into a playing field at the Insein Engineering Institute.

Fortunately, the village carpenter and tailor managed to repair the wing and she flew on to Allahabad and then to Calcutta. En route to Rangoon came the terror of a monsoon storm. Torrential rain and howling winds tried to wrench *Jason* out of the sky. Soaked to the skin and blinded by the rain, Amy searched desperately for the racecourse where she was to land at Rangoon. A temporary break in the storm and clouds gave her a glimpse of the field below, and immediately the Moth was spiralling down to safety. Suddenly Amy realized that this was not the racecourse, but it was too late to change her approach. Within seconds the aircraft had nosed over, smashing the propeller and damaging a wing. Fortunately she had landed on the playing field of

Amy's daring flight attracted much publicity and was used to advertise many products.

Insein Engineering Institute, some 16 kilometres (10 miles) from Rangoon. Willing hands speedily repaired the damage and Amy's spare propeller saved the day. It would have been impossible to take off from that field, so *Jason* was towed to Rangoon racecourse for take-off.

Amy flew on . . . from Rangoon to Don Muang, just short of Bangkok; on to Singora, short of Singapore; and then to Singapore, Semarang, and Sourabaya, all the time struggling against monsoon rains and buffeting winds. The repairs at Insein had dashed any hopes of beating Hinkler's record. Now it was Amy and *Jason* versus the elements, determined to finish the course. Atamboea, an island of the Timor group, was the next landing point. However, Amy was forced by winds to land in near darkness at Haliloelik, 19 kilometres (12 miles) short of her target. Finally she gained Atamboea, the last stepping stone towards her goal. On the morning of 24th May, Amy and her faithful *Jason* took off on the stage that she had dreaded most, the crossing of the Timor Sea to Port Darwin.

Then, suddenly, there it was—big, beautiful Australia, barely visible through Amy's tears of relief and excitement.

Map to show the route Amy took on her England-Australia flight: 1 Croydon; 2 Vienna; 3 Belgrade; 4 Constantinople; 5 Aleppo; 6 Baghdad; 7 Basra; 8 Bushire; 9 Bandar Abbas; 10 Karachi; 11 Jhansi; 12 Calcutta; 13 Rangoon; 14 Bangkok; 15 Singora; 16 Singapore; 17 Semarang; 18 Sourabaya; 19 Bima; 20 Kupang; 21 Port Darwin.

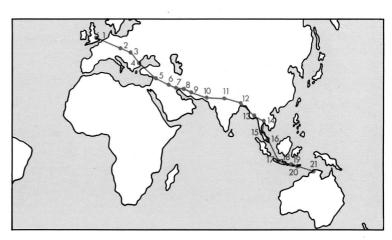

Amy takes off for the last time

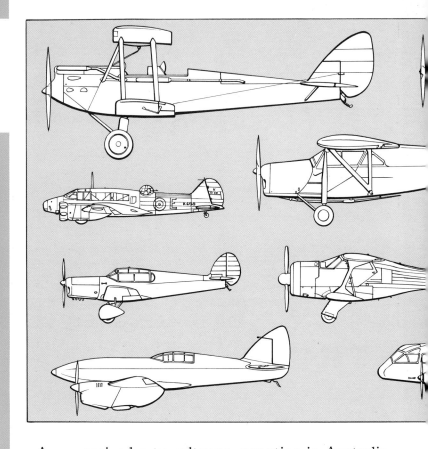

Amy acclaimed as world heroine... Honoured around the world... Campaigns for women to become involved in aviation ... First attempt at solo flight to Peking fails... Sets off again on London-Tokyo flight... Beats London-Cape Town record... London to Melbourne air race... Last record attempt... Second World War ... Amy joins the ATA... Final journey.

Amy received a tumultuous reception in Australia. The people responded generously to the courage of this young woman from faraway Britain. Bert Hinkler had taken three and a half days less to complete the journey, but it should be remembered that he was an experienced pilot. Before her Australian adventure, Amy's longest flight had been only 320 kilometres (200 miles). Her reception at Croydon, on returning to Britain, was a highlight of Amy's life. Nearly 200,000 people had gathered to greet her. As she stepped out of the Imperial Airways airliner that had carried her from Alexandria, their cheers were almost deafening. Honours and gifts followed, including an award of the C.B.E. from His Majesty the King, and £10,000 from the *Daily Mail*.

But Amy regarded this as only the beginning of her

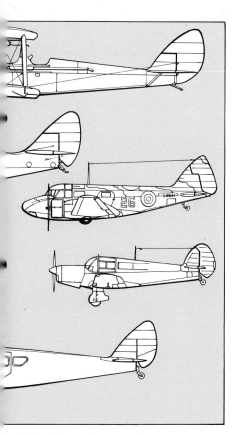

Profiles of some of the aircraft that Amy flew during her career.

career. She was an ardent campaigner for the advance of aviation, and for women to be part of its expansion. Amy took off again on 1st January 1931, attempting a solo flight to Peking, via Poland and Siberia. This time she was defeated by bad weather, and her Gipsy Moth *Jason III* was severely damaged while landing in fog north of Warsaw. Undaunted, she set off again on 28th July 1931 in a de Havilland Puss Moth, *Jason II*, accompanied by C. H. Humphreys, formerly the chief engineer of the London Aeroplane Club. Their aim was to beat the London-Tokyo record of seven days, but their final time for the flight was 8 days 22 hours 35 minutes. After considerable delays, due to bad weather, Amy landed back at Lympne, Kent, on 9th September.

Amy marries

In July 1932 Amy married Scotsman J. A. (Jim) Mollison, the pilot who had set a new Australia-England solo record in 1931. Only four months later, on 14th November, she took off from Lympne in a

Amy, Jim Mollison and Amy's mother are seen here being photographed by her father.

Portrait of Amy in full flying kit.

Puss Moth, *Desert Cloud,* in an attempt to establish a new London-Cape Town record. She not only broke the record, but also set a new Cape Town-London record on her return flight.

On 24th July 1935 during a flight from London to New York, both Amy and her husband were slightly injured when their de Havilland Dragon I *Seafarer* overturned on landing at Bridgeport, Connecticut. The big event of 1934 was the London-Melbourne air race. The Mollisons, flying a de Havilland Comet *Black Magic* were leading at Karachi, but had to withdraw with engine trouble at Allahabad.

It was not until 4th May 1936 that Amy set out on her last record attempt, hoping to regain the London-Cape Town and return records which had been taken by Flight Lieutenant Tommy Rose three months earlier. Flying a Percival Gull she achieved both her aims.

The next four years were perhaps the most frustrating of Amy's life. The days of pioneering flights were coming to an end; life was rather humdrum and routine. From 1938 thoughts of war were in most people's minds. When war finally came, Amy could not immediately see how best to serve her country. Then in the summer of 1940, she decided to apply to serve in the A.T.A. There was one thing she could do at least as well as any man, and that was to fly light aircraft. In the A.T.A. she found new happiness doing an important job with her usual efficiency and determination. Then came that fateful day of 5th January 1941, when Amy Johnson, only thirty-eight years old, gave her life for her country.

Amy at Croydon aerodrome after setting a new record for the solo flight from Cape Town to England.

Dates and events

- 1903 Amy Johnson born in Hull, England on 1st July.
- 1929 Amy makes her first solo flight (early June) and gains the private pilot's 'A' licence (late June).
- 1930 Amy becomes the first woman to fly solo from England to Australia (5th to 24th May).
- 1931 Takes off on unsuccessful flight to Peking (1st January).
- 1931 Makes successful England to Tokyo flight accompanied by C. H. Humphreys (28th July to 5th August).
- 1932 Marries J. A. Mollison (July).
- 1932 Establishes new solo flight record from England to Cape Town, South Africa (14th-18th November). On return flight establishes new solo flight record from Cape Town to England (11th-18th December).
- 1933 Amy and her husband make an unsuccessful attempt to fly from England to New York (22nd-24th July).
- 1934 Takes part with her husband in MacRobertson England-Australia air race but forced to retire at Karachi (20th October).
- 1936 Sets new solo flight record from England to Cape Town (4th-7th May).
- 1936 Sets new solo flight record from Cape Town to England (10th-15th May).
- 1941 Amy Johnson dies while delivering an aircraft in service with the A.T.A.

Jean Batten

The early 1920s were days of great excitement in the aviation world. In New Zealand a ten-year-old girl dreamed of flying around the world. Leaving her native land, Jean Batten went to England to learn all she could about flying. She became determined to fly from England to Australia. Twice she tried . . . twice she failed. At the third attempt she succeeded, arriving home to a hero's welcome. She had fulfilled her childhood dream.

From an early age Jean Batten met travellers from all over the world. Her parents enjoyed entertaining many of the visitors who came to see the natural wonders of Rotorua, New Zealand, where Jean was born on 15th September 1909. Just before her fifth birthday her father sailed to Europe. He was one of the many volunteers from New Zealand who left the security of their distant islands to fight alongside the British and their allies during the First World War. His safe return from the other side of the world convinced Jean that world travel was not particularly difficult.

The war had done much to develop the somewhat frail aeroplanes in use at its beginning. When peace returned, pioneer aviators began to look out across the oceans. In 1919 came the first non-stop flight across the Atlantic, and later that year Keith and Ross Smith linked England and Australia by air for the first time. These were exciting events, especially for New Zealanders. Their isolated position in the Tasman Sea, more than 1,610 kilometres (1,000 miles) from the nearest major landmass of Australia, made long distance air travel of great interest.

For ten-year-old Jean it was a thrilling prospect: perhaps one day she might be able to fly across the world. Then, in 1928, Charles Kingsford Smith (later

Jean's private pilot's 'A' licence dated 5th December 1930.

Jean spent many months preparing her Gipsy Moth for the England-Australia flight.

Sir Charles) and a crew of three made the first full crossing of the Pacific Ocean by air. Three months later he captained *Southern Cross* again to link Australia and New Zealand by air for the first time. Just before her nineteenth birthday Jean was introduced to this handsome hero. Her mind was made up there and then: she would become a pilot.

Her first problem was her father, who thought it far too dangerous. He also pointed out that it would be very expensive. Jean did not think either of these were valid reasons to stop her from flying. It would not be dangerous if she was properly trained, and she could sell her piano to help pay for flying lessons.

Jean goes to London

In 1929 Jean arrived in Britain and joined the London Aeroplane Club, where Amy Johnson was just completing her flying training. Like Amy she had the ambition to fly from England to Australia, but after gaining her private pilot's 'A' licence, she could not find backing for such a flight and returned to New Zealand. She thought it might be easier to raise support in her home country, but soon realized that her very limited flying experience was unlikely to enthuse anyone.

So she returned to England in 1931, determined to gain the commercial 'B' licence, which demanded at least 100 hours' flight experience. In addition to the other subjects required for this licence, one of which was navigation, she decided to study the basics of airframe and engine maintenance. It was a difficult period and, like Amy Johnson, she had to save every penny to enable her to complete her training. Even when Jean succeeded in gaining her licence, she still could not find a sponsor for the flight.

Her luck changed in 1932, when another pilot of the London Aeroplane Club agreed to help her with costs. At last she could make plans for the flight.

Third time lucky for Jean

Take-off on England-Australia flight . . . Mechanical troubles cause forced landing . . . Second attempt ends in failure . . . Third attempt looks more promising . . . Spends night in the desert due to sandstorm ahead . . . Oil leak develops . . . Flies through monsoon storm . . . Thick fog threatens to delay next stage . . . Final stage across Timor Sea . . . Successful landing.

On 9th April 1933, after what seemed to be endless months of preparation, the Gipsy Moth stood ready for the England-Australia record attempt at Lympne airfield, Kent. The engine of the little plane responded with a roar as Jean pushed the throttle lever wide open. Slowly the heavily-laden aircraft began its take-off run across the airfield. Soon daylight showed beneath the wheels, and Jean could relax and set course on the first stage of the long route which lay ahead. The journey was to end in disappointment, however, for eight days later the engine died as she was approaching Karachi. Jean made a forced landing some 5 kilometres (3 miles) short of the airfield. The aircraft was damaged, but fortunately she escaped injury.

Just over a year later, on 22nd April 1934, Jean was ready to make a second attempt. This time the aircraft was her own—a Gipsy Moth which she had bought for £260. However, a lot more money had to be spent overhauling the engine and airframe before it was ready for the flight to Australia. This attempt was even less successful, ending at Rome when, in

Jean turns over the propeller of her Gipsy Moth to start the engine.

After months of preparation and two abortive attempts at the England-Australia flight, Jean was finally well on her way.

darkness and torrential rain, the engine spluttered to a stop. Out of petrol after battling against headwinds between Marseilles and Rome, Jean managed to make a near-perfect power-off landing in a tiny field. Miraculously she avoided high wireless masts and cables that were invisible in the darkness. The Gipsy Moth came to a halt within yards of the River Tiber.

Lesser characters might have considered this narrow escape more than enough. But not Jean Batten who, by May 6th 1934, had flown her aircraft back to England. Two days later she began her third attempt. The intended route lay via the Mediterranean shores to Cyprus, on to Syria, Iraq, Persia (now Iran), India, Burma, Malaya (now Malaysia), Singapore, then along the string of islands that then formed the Dutch East Indies, and finally from Kupang across the Timor Sea to Darwin in Australia.

Most of the journey went well, but as she approached Baghdad a violent sandstorm raised a dust haze which completely obscured her flight path. She had to turn back and spend the night at Fort Rutbah in the Syrian Desert. There, camels and an Arab encampment provided a strange contrast with her Gipsy Moth. The only worrying mechanical problem occurred between Allahabad and Calcutta, when an oil leak developed. This caused Jean great anxiety, for it would have been impossible to make a forced landing on the hilly, wooded country she was flying over. Fortunately, the engine kept running, but an oily black stain almost covered the Moth by the time of touch-down at Calcutta.

Far more frightening, however, was the monsoon storm that barred her approach to Victoria Point at the most southerly tip of Burma. The thunder of rain battering against the wing and fuselage almost drowned the noise of the engine—it seemed a miracle that it kept turning. Visibility was so bad that Jean could not see the wingtips from the cockpit, which by then was almost flooded. Although it should have been light, the storm cloud had turned day into night. Luminous instruments on the dashboard glowed with an eerie green light. Soaked to the skin, despite her tropical flying suit, Jean spent almost forty minutes

Near Baghdad Jean hit a violent sandstorm which forced her to land at Fort Rutbah in the Syrian Desert.

Map to show the route Jean took on her England-Australia flight: 1 Lympne; 2 Marseilles; 3 Rome; 4 Brindisi; 5 Athens; 6 Nicosia; 7 Damascus; 8 Baghdad; 9 Basra; 10 Bushire; 11 Jask; 12 Karachi; 13 Jodhpur; 14 Allahabad; 15 Calcutta; 16 Akyab; 17 Rangoon; 18 Victoria Point; 19 Alor Star; 20 Singapore; 21 Batavia; 22 Sourabaya; 23 Rambang; 24 Kupang; 25 Port Darwin.

struggling against the storm and, at the same time, trying to establish her position. Suddenly, far below, she caught a glimpse of the airfield she was seeking, and quickly banked the little aircraft down through a hole in the cloud mass. The airfield at Victoria Point looked little more than a lake, and the Gipsy Moth threw up streams of spray as Jean landed.

Final leg on the journey home

At Batavia thick fog threatened to delay the flight. To make a take-off possible, the local fuel agent drove his car back and forth across the airfield to clear a pathway through the fog. This made it just possible for Jean to attempt a take-off, climbing blindly as soon as the wheels left the ground. Imagine her relief as the aircraft broke through the top of the fog into the sunlight above.

Soon the last daunting step was ahead, the endless hours of flight across the featureless Timor Sea. Although the target—Australia—was comfortingly large, there were times towards the end of the crossing when Jean wondered if she had completely missed this vast landmass. Then, suddenly, after nearly seven hours of flight, there was land, growing bigger and more beautiful every second.

As the wheels of the Moth touched the airfield at Darwin on 23rd May 1934, Jean Batten had completed her solo journey in 14 days 22 hours and 30 minutes, beating by more than four days the time established by Amy Johnson.

The warm-hearted Australians cheered in their thousands when she arrived later at Sydney. But this tumultuous reception was nothing compared with the welcome she received when she steamed into Auckland harbour aboard the SS *Aorangi*. Ships at anchor sounded their sirens in salute, massed bands played, and the thousands who stood or clung to every vantage point roared their greeting. New Zealand's heroine had come home.

First woman to fly across the South Atlantic

Leisurely flight back to England... One scary moment as engine splutters to virtual standstill mid-flight... Lands at Croydon... Plans to become first woman to fly across the South Atlantic... Take-off from Lympne... Bad weather... Engine-off landing in Senegal... Hair-raising take-off on next stage... Heavy rain and thick cloud obscure the route ahead... Beats previous record for the flight.

After the excitement of her homecoming, and a six-week triumphal tour of New Zealand, Jean soon began making plans for future conquests. At the top of her list was an attempt to become the first woman to fly across the South Atlantic Ocean. The problem was that the Gipsy Moth needed to be returned to Britain. Should she send it by sea? No! Why not fly back to England?

On 12th April 1935 Jean took off from Darwin on the return Australia-England flight, and was soon to have the most frightening experience of her life. About 400 kilometres (250 miles) out over the Timor Sea, flying at an altitude of about 1,830 metres (6,000 feet), the engine of the Moth suddenly spluttered and virtually stopped, the propeller just turning very slowly. A quick glance around the cockpit showed no cause for this, and helplessly Jean put the aircraft into a shallow glide. The only sound as the Moth dropped lower and lower was the wind blowing through the rigging wires. With less than 305 metres (1,000 feet) to the surface of the waves, Jean was convinced that this was the end of everything, when suddenly the engine burst into life again. With a great sigh of relief she was able to regain height and continue her journey.

Welcome home for Jean

Fortunately, this was the only serious problem during her unhurried return flight. When she landed at Croydon on 29th April, Jean had become the first woman to complete a two-way solo flight between England and Australia.

Preparations for the South Atlantic flight included the replacement of the Gipsy Moth by a new Percival Gull monoplane, and at 06.30 hours on 11th November 1935 Jean took off from Lympne, Kent, towards her first planned stop at Casablanca, Morocco. This stage of the journey was plagued by

A scary moment for Jean on the return Australia-England flight as the engine of her Gipsy Moth comes to a virtual standstill over the Timor Sea.

bad weather: clouds, rain and hailstorms. Over the Bay of Biscay she had to climb above 4,270 metres (14,000 feet) to clear the clouds, before crossing the Pyrenees. Without further problems Jean landed at Casablanca, at 16.15 hours after a non-stop flight of 2,173 kilometres (1,350 miles).

Less than twelve hours later the Gull was airborne again, heading for a refuelling point near Villa Cisneros (now Dakhla) on the coast of Mauritania. Jean wasted no time there as another 1,095 kilometres (680 miles) lay ahead of her that day before she would reach Thies, Senegal, the jump-off point for the Atlantic crossing. The landing there was rather hair-raising, the throttle control jamming so that the engine would not run at low speed. In desperation Jean had to make her approach, switch off the engine completely, and make a tricky engine-off landing. The subsequent take-off was even more frightening, for only a small area of the airfield was in use. It was still dark and raining heavily when Jean positioned the Gull as near as possible to one corner

Jean relaxes at home with just some of the messages of congratulations she received.

of a diagonal. The moment for take-off had come, but the little aircraft was gaining speed so slowly that it seemed it would never leave the ground. It was not until she had crossed almost the entire airfield that Jean managed to coax the plane into the air, barely skimming the treetops as she headed toward Dakar, and then out across the Atlantic to a landfall in Brazil almost 3,220 kilometres (2,000 miles) away.

Torrential rain and cloud taxed Jean's navigational skill to the utmost. Even more, the 13 hour 15 minute flight demanded all her courage to overcome the loneliness of the tiny enclosed cabin. With no radio with which to seek help or advice, she had only her own ability to ensure success. There was great

Jean's Percival Gull barely skimmed the treetops when she took off from Thies, Senegal.

excitement when she made landfall less than a mile off course, circled Natal, and taxied towards the crowds that were gathered to greet her. She was even more excited when it was confirmed that her total flight time was 2 days 13 hours 15 minutes beating Jim Mollison's previous record by almost a day.

Before leaving Brazil, Jean decided not to miss the opportunity of seeing something of South America. She flew to Rio de Janeiro, Buenos Aires, and Montevideo, and was honoured at every stopping point. Almost reluctantly, Jean and her precious Gull boarded the SS *Asturias* at Buenos Aires on 6th December, arriving at Southampton just before Christmas.

After crossing the South Atlantic, Jean flew her Percival Gull to several cities along the South American coast.

New Zealand's heroine sets more records

Final ambition to fly solo from London to New Zealand . . . Take-off from Lympne . . . Punctured tyre threatens to end flight . . . Emergency repair saves the day . . . Lands at Darwin breaking previous record for England-Australia flight . . . Safe landing before huge crowds at Auckland . . . On return flight to England sets new record for Australia-England flight.

Jean's burning ambition ever since she learned to fly had been to complete a solo flight from London to New Zealand. Planning for this kept her occupied throughout the summer of 1936, and at 04.10 hours on 5th October 1936 she took off from Lympne, Kent. The experience and confidence gained from earlier record-breaking flights made the flight much easier. Airfields and ground facilities were also beginning to improve.

The most serious difficulties were due to the weather. Fortunately, in most cases the Gull could climb above dust or rainstorms. However, a punctured tyre at Kupang threatened to end the flight. Jean had no spare inner tube, nor was one available in that remote area. Then someone had the bright idea of stuffing the tyre with rubber sponges. This emergency repair proved good enough for the take-off, and after an easy and fast crossing of the Timor Sea, Jean landed at Darwin just 5 days 21 hours 3 minutes after leaving Lympne. It was not only a new England-Australia solo record, but it had shattered the previous record held by Harry Broadbent by 24 hours 16 minutes.

Just over two days later she had crossed the vast continent of Australia from Darwin to Sydney. Jean was delayed there for two days, waiting for suitable weather to tackle the Tasman Sea. At 04.37 hours on 16th October she lifted the little Gull off the military airfield at Richmond, some 64 kilometres (40 miles) from Sydney. Ahead lay a daunting 2,410 kilometres (1,330 miles) of ocean before she would reach New Plymouth on the west coast of New Zealand's North Island.

This air crossing was considered so dangerous that immediately before take-off, Jean instructed the Station Commander at Richmond: 'If I go down in the sea no one must fly out to look for me . . . I have no wish to imperil the lives of others . . .' This was typical of Jean's consideration and courage and,

happily, no one needed to make such a search.

Making her landfall almost exactly on course, she flew the extra 248 kilometres (154 miles) to land before huge crowds at Auckland. She had indeed earned their welcome, completing the first solo and direct flight between England and New Zealand, in 11 days 1 hour 25 minutes.

Most of her friends imagined that this would be sufficient to satisfy Jean's burning desire to fly, and be first. They were, of course, wrong. Taking off from Darwin on 19th October 1937, she set out on an attempt to capture the solo record for an Australia-England flight. Like the outward journey to Australia the previous year, this became a battle against the weather. More than anything else it needed all her reserves of courage to keep pushing ahead despite the weather, and no matter how exhausted she felt.

Jean Batten and Judith Chisholm are pictured here together after Judith had broken Jean's 44-year-old record for the flight from England to New Zealand.

Profile of Jean's Percival Gull. The inset map shows the routes she took on her record-breaking flights to South America and New Zealand.

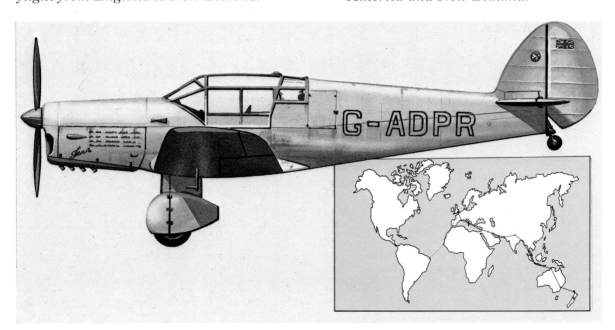

Her reward, when she landed at Lympne on 24th October, was a new Australia-England record of 5 days 18 hours 15 minutes, beating the existing record by 4 hours 10 minutes. She had become the first person to hold both the Australia-England and England-Australia solo records at the same time. 'And she's only a woman' said one of the newspaper reports. She had surely proved once and for all that women were no longer the weaker sex.

Jean Batten continues to lead an active life, the only one of the four courageous women in this book who is still alive. Her interest in aviation has lasted through the years, and she can look back with pride on her own contribution to the aviation scene. Her achievements, and those of other pioneer pilots, both male and female, helped speed the development of air routes and aircraft which now carry most of the world's long distance passengers.

Dates and events

1909 Jean Batten born at Rotorua, New Zealand on 15th September.
1929 Jean makes her first solo flight.
1933 Makes her first, unsuccessful, attempt to fly solo from England to Australia (9th April).
1934 Makes her second, unsuccessful, attempt to fly solo England-Australia (22nd April).
1934 At third attempt establishes new women's solo record England-Australia flight (8th-23rd May).
1935 Jean flies from England to Brazil, establishing a new record for the flight and a new record speed for the crossing of the South Atlantic, and becomes the first woman to have flown the South Atlantic (11th-13th November).
1936 Jean flies from England to New Zealand, setting a new solo flight time for England-Australia flight. Makes the fastest solo crossing of the Timor Sea as well as the first solo flight from England to New Zealand (5th-16th October).
1937 Jean establishes a new solo record flight from Australia to England (18th-24th October).
1980 Welcomes in Auckland, New Zealand, British pilot Judith Chisholm after she had broken Jean's 44-year-old record for England-New Zealand flight (25th November).

Jacqueline Cochran

> Jacqueline Cochran received little education and got her first job at the age of eight. However, by the time she died she was world-famous. How did she manage to achieve such fame? Perhaps her greatest quality was her willingness to work hard. As well as competing in the male-dominated field of aviation, she also established her own successful cosmetics business. She proved that by hard work and determination, the sky really was the limit.

On a bright, clear day in May 1953, the American Jacqueline Cochran found herself at the controls of a Canadian-built version of the North American F86 Sabre, the first swept-wing, turbojet-powered fighter to enter service with the United States Air Force. This was the moment she had been waiting for ever since her first flight in a jet-powered aircraft nine years earlier. Instead of just flying this fighter, she was determined to establish a new record.

Flying at a height of about 13,715 metres (45,000 feet), Jackie nosed the Sabre over into a full-power, near-vertical dive. The high-pitched roar of the engine filled the sky, but Jackie could hear little in the cockpit of the fighter, as it dropped towards the earth far faster than any stone. As the speed built up, her eyes flickered between the altimeter, which showed that she was losing height at an almost frightening rate, and the Mach meter, which indicated the aircraft's speed as a percentage of the speed of sound. As the speed of the plane increased, she called the readings over the radio to Charles ('Chuck') Yeager, who was flying overhead. He was the first man in the world to pilot an aircraft faster than the speed of sound. His job was to keep an eye on Jackie, and instruct her to reduce speed, if he thought she was running into trouble.

In May 1953, Jackie put her Sabre into a near-vertical dive and became the first woman to fly faster than the speed of sound.

'Mach .90' (90 per cent of the speed of sound), she called; then 'Mach .95'. The speed was still building up rapidly. 'Mach .97, Mach .98'. Suddenly the aircraft began to shudder as shock waves of air built up ahead of the fuselage and wings. 'Mach .99, Mach 1.0, Mach 1.01'. As she eased back on the power control and gently, oh so gently, began to pull out of the dive, her earphones came alive. 'Congratulations, Jackie, you've made it.' These words from Chuck Yeager came as music to her ears. She had become the first woman in the world to fly an aircraft faster than the speed of sound. Gradually the speed dropped, and there was the same shuddering and instability as she slipped back through the sound barrier. A wing dipped for a second, but was quickly corrected, and Jackie headed the Sabre back to land and to a host of congratulations.

From rags to riches: Jackie's early life

Early life of poverty ... Little chance to receive any education ... First job in a cotton mill ... Becomes involved in hairdressing and beauty world ... First flying lessons ... Shows natural talent for flying ... Receives pilot's licence and makes first long-distance flight ... Determines to make her life in flying ... Works as an unpaid air stewardess in return for flying instruction.

Like the other pilots in this book, Jackie Cochran was determined to succeed. But whereas the other three each received help from their parents, Jackie had only her own courage to drive her on.

Brought up by foster parents, and never knowing her real parents, her early life was one of real poverty. Until the age of seven or eight, she had not worn either shoes or stockings, and her dresses were made from old flour sacks. All too often there was nothing to eat; fish and beans were a real meal, and 'luxuries' like sugar and butter were unheard of.

There was no chance of Jackie receiving any education. She learned her alphabet from the signs written on the rolling stock of freight trains that steamed noisily past the lumber camp where she lived. Beyond knowing that she was born in 1906, she had no idea when her birthday was and so she chose her own.

Jackie's first job

Later she met a schoolteacher who taught her to read, gave her her first real dress, and helped her to understand the importance of cleanliness. At the age of eight Jackie, still shoeless, got her first job, in a cotton mill. The first things she bought when she had enough money were two pairs of shoes. Two years later she was put in charge of fifteen other children, but soon after was forced to leave when the mills went on strike for three months.

This was the turning point in Jackie's life. She found work in a ladies' hairdressing salon in Columbus, Georgia. From there she moved to a beauty salon in Montgomery, Alabama, where she saved enough money to buy a secondhand Model T Ford car. With this she began to travel around the country, selling dress patterns. Although this work was interesting and profitable, she decided to return to the hairdressing and beauty salon business. Once

Vincent Bendix congratulates Jackie after she had won the 1938 Bendix Trophy Race.

again her keenness to work quickly brought success.

During this period Floyd Odlum, who was later to become Jackie's husband, suggested that she should learn to fly. Thus, in 1932, she received her first flying lessons. It was immediately clear that she was a natural pilot: within three days she had flown solo; she received her flying licence within three weeks; and she made her first long-distance flight to Canada very soon after that. Badly bitten by the aviation bug, Jackie decided that, somehow, she must find ways and means of earning her living in aviation. The most important thing was to gain as much flying experience as possible, so that she would be at least as good as male pilots. Working as an unpaid stewardess for an airline, in return for the occasional chance of taking over the controls, she received first-class training, especially in instrument flying. The moment had come when Jackie felt she was ready to compete in the man's world of aviation.

Jackie hoped to win the 1934 England-Australia air race in this special version of a Gee Bee racer.

Jackie combines business and aviation

Jackie gains experience in competitive flying... Establishes her own cosmetics business... Shows flair for business... Enters England-Australia race... Forced to retire from race due to mechanical problems... Wins the Bendix Trophy race at second attempt... Sets new transcontinental speed record... More records... Proves that she can compete equally with male pilots.

For Jackie aviation not only had to be interesting—it had to be profitable too. One area which promised high rewards was aircraft racing, but if she was to win prizes of any value, it meant competing against men, and not women. To get some experience of what might be involved, Jackie took part in an all-women's race at Roosevelt Field, Long Island, New York, in the autumn of 1933.

The year 1934 was a busy one for Jackie. As well as entering competitive flying on a serious basis, she also established her own cosmetics business. The major event in the aviation world that year was the England-Australia air race from Mildenhall, Suffolk, to Melbourne. This race was part of Melbourne's centenary celebrations. First prize in the speed section was £10,000, which was then a great deal of

Jackie flew this Northrop Gamma aircraft in the 1935 Bendix Trophy Race.

Profile of Jackie's Northrop Gamma.

Jackie, flying a modified Seversky P-35 single-seat fighter, won the 1938 Bendix Trophy air race.

money. Jackie felt confident that she could win. She chose as partner an airmail pilot, Wesley Smith, who had given her intensive training in instrument flight. Jackie thought a special version of one of the Granville Brothers Gee Bee racers would be fast enough to enable them to win without too much difficulty. They were, however, forced to retire at Bucharest, Romania, after mechanical problems had developed. There had just not been enough time to get all the bugs out of the various systems of the newly-constructed aircraft.

In 1935 Jackie entered the Bendix Trophy Race, a U.S. national long-distance event. As she took off in

Jackie relaxes alongside her Seversky fighter. This aeroplane was developed in the U.S.A. by the company founded by ex-Russian Major Alexander P. de Seversky.

foggy conditions at 03.00 hours on 2nd September, the engine of her Northrop Gamma aircraft was low on power. Having started her take-off run, and unable to see the ground ahead, she had somehow to lift the machine into the air. At the end of the field she was so close to the ground that her trailing radio aerial was torn off by the perimeter fence. Fortunately she was able to climb slowly up over the sea until she could see the stars, but on this occasion too she had to retire from the race.

In 1938 she succeeded in winning the Bendix Trophy. After receiving the award at Cleveland, Ohio, Jackie climbed back into her aircraft and flew on over Pennsylvania and New Jersey to New York to establish a new transcontinental speed record. She was to prove by hard work that a woman could compete equally with men in the world of competitive flying. At one time Jackie held more than 200 U.S. national records.

Jackie regarded one record in particular as a highlight of her career. This was a new world speed record over a 100 kilometre closed circuit. Closed circuit means that the course is circular. The circuit is marked out by twelve pylons, and the aircraft has to be flown outside these pylons at a height of only 91 metres (300 feet). Observers at each pylon check that the aircraft is flown on track, and electronic equipment provides split-second timing. The aircraft has to be flown with great accuracy around the circle, just clearing the pylons. The slightest error can lose a record or transform the aircraft into a blazing stream of wreckage. There is one other problem: because fuel is heavy, only the absolute minimum is carried. For her record attempt Jackie had surplus fuel for only two minutes' extra flight. This meant that she had to follow her set plan exactly, otherwise there would be the complication of landing with a dead engine.

One day in May 1953, 47-year-old Jackie flew a Sabre turbojet-powered fighter around this course at a speed of just over 1,049 km/h (652 mph), gaining the world speed record for an aircraft of this class. She had captured the record from Colonel Fred Ascani by a margin of 24 km/h (15 mph). He was among the first to send his congratulations on this great achievement, commenting '. . . you have accomplished the impossible and have shown what . . . determination can do'.

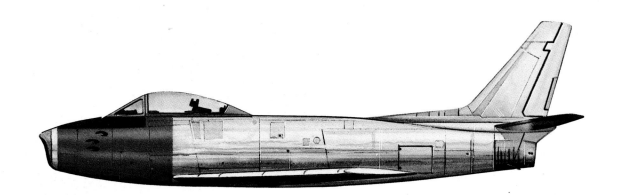

Profile of a Sabre turbojet-powered fighter.

World acclaim for Jackie

Jean Batten, Amelia Earhart and Amy Johnson each made their contribution to aviation in its early years between the First and Second World Wars, long before ordinary men and women had come to accept the aeroplane as an everyday means of travel. Now it not only speeds the businessman on his way, but enables people like you and me to enjoy holidays in faraway places.

Jacqueline Cochran was not one of these trailblazers, but she showed that women could play an increasingly important part in the aviation scene. During the Second World War she recruited a group of twenty-five women pilots and travelled with them to Britain to serve in the Air Transport Auxiliary (A.T.A.), ferrying aircraft from manufacturers to operational units. Serving in the A.T.A. as a flight captain, she returned home when America became involved in the war. Here she was to play a vital role in the creation of the Women's Airforce Service Pilots (W.A.S.P.s) organization. These women not

Jackie encourages women to become involved in aviation... Recruits groups of women pilots to serve in the A.T.A. during the war... Returns home to America... Becomes Director of W.A.S.P.s... Covers closing stages of war as correspondent for Liberty magazine... Selected as America's outstanding businesswoman... Appointed President of the International Aeronautical Federation.

Jackie signs autograph books for women pilots from the W.A.S.P.s organization.

Among the aircraft Jackie flew was this version of the Mustang.

only ferried aircraft, but were also used for duties such as routine flight testing and target towing.

After being Director of W.A.S.P., and with the war nearing its end, Jackie talked herself into becoming a correspondent for *Liberty* magazine. Sent out to cover the closing stages of the war in the Pacific, she witnessed the formal surrender proceedings in the Philippines of Japan's General Yamashita, and later flew over Tokyo and saw the devastation caused by incendiary bomb attacks.

Pursuing her interest in so many different aspects

In May 1963 Jackie piloted this Lockheed TF-104G Super Starfighter to a new 100-km closed circuit world speed record.

of aviation, Jackie travelled to almost every part of the world. The winner of numerous trophies, she also received a large number of honours and awards, including the United States Distinguished Service Medal, the Harmon International Trophy, and the Gold Medal of the Fédération Aéronautique Internationale, the organization which governs aviation sport activities on an international basis, including aviation world records.

Jacqueline Cochran, a lady of remarkable achievements, died on 8th August 1980 at the age of seventy-four. A guest of presidents and monarchs, she must have found it difficult to decide which of two honours meant most to her: selection as America's outstanding businesswoman, or her appointment as President of the Fédération Aéronautique Internationale.

Writing from her own experience of rags to riches, she once penned a few words to young people. Thinking of them as the aeroplanes, which had been so much a part of her life, she concluded, 'If you will ... pour in the fuel of work and still more work, you will be likely to go places and do things.' It had certainly been true for Jackie!

Dates and events

1906 Jacqueline Cochran born in Florida, U.S.A.
1932 Jackie makes her first solo flight.
1934 Takes part in MacRobertson England-Australia race, but retires at Bucharest, Romania (20th October).
1935 Takes part in Bendix Trophy Race, but retires after hazardous take-off in fog (2nd September).
1938 Jackie wins Bendix Trophy Race and after receiving trophy flies on to establish a new U.S. transcontinental record flight time.
1940 Establishes new 2,000-km closed circuit world speed record (April).
1945 Jackie is awarded U.S.A.A.F. Distinguished Service Medal (21st May).
1946 Jackie placed second in Bendix Trophy Race.
1947 Establishes a new 100-km closed circuit world speed record.
1953 Jackie becomes the first woman to fly faster than the speed of sound (May). Establishes a new 100-km closed circuit world speed record (May).
1980 Jackie Cochran dies (8th August).

Glossary

'A' licence A private pilot's licence, which is awarded after a pilot has proved his or her ability to fly an aircraft.

Altimeter An instrument to measure the height above sea level at which an aircraft is flying.

Altitude The height above sea level at which an aircraft flies.

Biplane An aeroplane with two wings mounted on each side of the fuselage.

'B' licence A professional pilot's licence, which must be held by all pilots who fly commercial transport aircraft.

Cabin An enclosure where the aircraft's occupants sit.

Cockpit An unenclosed space where the pilot and sometimes another occupant sit.

Engine-off landing A landing made without using any engine power during the approach and final stages.

Flight path The invisible track through the air taken by an aircraft as it flies from one place to another.

Fuselage The main body structure of an aeroplane.

Instrument flight When bad visibility or darkness make it impossible for the pilot to see outside the aircraft, he or she has to rely on instruments to maintain level flight.

Landfall First sight of land after making an over-water flight.

Monoplane An aeroplane with a single wing on each side of the fuselage.

Pioneering flight The first flight over a new route or the first flight in a new type of aircraft.

Solo flight A flight carried out by a pilot, without any other person being on board the aircraft.

Throttle A control which allows the pilot to vary the power output of an aircraft's engine.

Further reading

The Stars at Noon by Jacqueline Cochran and Floyd Odlum (Arno Press, 1979 reprint of 1954 edition)

Heroines of the Sky by Jean Adams et al. (Arno Press, 1942)

The Sky's the Limit: Women Pioneers in Aviation by Wendy Boase (Macmillan, 1977)

The Fun of It by Amelia Earhart (Academy Chicago Ltd., 1977)

Skystars: The History of Women in Aviation by Ann Hodgman and Ruby Djabbaroff (Atheneum Publishers, 1981)

Heroines of the Sky: Women in Aviation by Herve Lauwick (Beachcomber Books, 1960)

Women Aloft by Valerie Moolman (Time-Life Books, 1981)

We Were Wasps by Winifred Wood and Dorothy Swain (Aviation Book Co., 1978)

Winged Legend: The Life of Amelia Earhart by John Burke (Berkley Publishers)

20 Hours, 40 Minutes: Our Flight in the Friendship edited by James Bilbert (Arno Press, reprint of 1928 edition)

Index

Air Transport Auxiliary
  (A.T.A.) 21, 23, 34, 59
Ascani, Colonel Fred 58
Atlantic crossings 5-7, 12-15,
  43-6

Balchen, Bernt 12-13
Batten, Jean 37-49
  childhood 37
  England to Australia flight
    38-42, 43, 47
  first interest in flying 37
  flight training 38
  London to New Zealand flight
    47-9
  record attempts 47-9
  solo flights 47-9
  South Atlantic crossing 43-6
Bendix Trophy Race 56-7
Brancker, Sir Sefton 26
Broadbent, Harry 47

Cochran, Jacqueline 51-61
  breaks sound barrier 51-2
  childhood 53
  education 53
  establishes cosmetics
    business 55
  first job 53
  flight training 54
  international recognition 61
  speed records 57-8

Desert Cloud 34

Earhart, Amelia 5-19
  childhood 9
  disappearance 19
  first interest in flying 9
  flight training 10
  marriage 11

medical studies 9
record attempts 10, 11, 16, 17
round-the-world flight 17-19
transatlantic crossings 5-7,
  12-15
England-Australia air race 55-6
England-Australia solo flights
  25-32, 38-42, 47

*Fédération Aéronautique
  Internationale* 61

Gordon, Louis 5
Gorski, Eddie 13
Guest, Mrs Frederick 5, 8

Hinkler, Bert 26, 29, 32
HMS *Berkeley* 23
HMS *Haslemere* 22-3

Jason 27-31, 33
Johnson, Amy 21-34, 38, 42
  childhood 24
  disappearance 21
  England to Australia flight
    25-32
  final journey 21-3
  first interest in flying 24
  flight training 24-5
  marriage 33
  record attempts 34
  solo flights 25-6, 27-31, 32, 33
Jorgensen, Anders 24

Lindbergh, Charles 8, 10, 13, 16
London Aeroplane Club 24, 27,
  33, 38
London-Melbourne air race 34
London to New Zealand flight
  47-9

Mollison, Jim 33

National Aeronautic
  Association 10
Noonan, Fred 18, 19

Odlum, Floyd 43

'Powder Puff Derby' 11
Putnam, George Palmer 8-11

Rose, Tommy 34

Seafarer 34
Second World War 21, 34, 59
Smith, Charles Kingsford 37-8
Smith, Keith and Ross 37
Smith, Wesley 56
South Atlantic crossing 43-6
Southern Cross 38
Spirit of St. Louis 13
Stultz, Wilmer 5-7, 8, 12

Wakefield, Lord 26
Women's Airforce Service Pilots
  (W.A.S.P.s) 59-60

Yeager, Charles 51-2

Picture acknowledgements

The publisher would like to thank all those who provided illustrations on the following pages: BBC Hulton Picture Library 26 (top), 35, 45 (top); Molly Jones 25 (both), 33, 34; London Express News Service 48 (top); David Mondey *front cover,* 8, 12, 19; Photri, U.S.A. 5, 10, 16; Robert Pooley 36; Popperfoto 7, 15 (bottom), 23; RAF Museum 38 (both), 39; John Topham Picture Library 31 (top); Underwood, Collinge & Associates 54 (top), 57, 59; Malcolm S. Walker 14; John W. Wood & Associates 4, 6, 9, 11, 13, 17, 18, 20, 22, 26 (bottom), 28, 29, 30, 31 (bottom), 32-3, 36 (inset), 40, 41, 42, 44, 45 (bottom), 46, 48 (bottom), 52 (both), 54 (bottom), 55, 56, 58, 60, 61.